Maintaining Military Medical Skills During Peacetime

Outlining and Assessing a New Approach

Christine Eibner

Prepared for the Office of the Secretary of Defense

Approved for public release; distribution unlimited

RAND

NATIONAL DEFENSE RESEARCH INSTITUTE and
RAND HEALTH

This research was sponsored by the Director of Program Analysis and Evaluation and the Assistant Secretary of Defense for Health Affairs. It was conducted jointly by the RAND Health Center for Military Health Policy Research and the Forces and Resources Policy Center of the RAND National Defense Research Institute. The latter is a federally funded research and development center sponsored by the Office of the Secretary of Defense, the Joint Staff, the Unified Combatant Commands, the Department of the Navy, the Marine Corps, the defense agencies, and the defense Intelligence Community under Contract W74V8H-06-C-0002.

Library of Congress Cataloging-in-Publication Data

978-0-8330-4291-0

Published 2008 by the RAND Corporation
1776 Main Street, P.O. Box 2138, Santa Monica, CA 90407-2138
1200 South Hayes Street, Arlington, VA 22202-5050
4570 Fifth Avenue, Suite 600, Pittsburgh, PA 15213-2665
RAND URL: http://www.rand.org/
To order RAND documents or to obtain additional information, contact
Distribution Services: Telephone: (310) 451-7002;
Fax: (310) 451-6915; Email: order@rand.org

Preface

This monograph assesses a hypothetical model for maintaining the operational readiness of military medical personnel by stationing a subset of these personnel in nonmilitary settings, such as civilian hospitals. It presents results from a targeted set of interviews with civilian health care organizations for the purpose of evaluating the feasibility of the proposed model from a civilian standpoint. In addition, it discusses steps that the U.S. Department of Defense (DoD) could take to implement a pilot study of the proposed model that would gauge the model's effect on readiness, retention, and morale.

We received approval from RAND's Human Subjects Protection Committee to conduct interviews for this project. As part of this agreement, we keep the identities of interviewees and the names of their employers confidential.

This research was sponsored by the Assistant Secretary of Defense for Health Affairs and the Office of Program Analysis and Evaluation. It was conducted jointly within the RAND Health Center for Military Health Policy Research and the Forces and Resources Policy Center of the RAND National Defense Research Institute, a federally funded research and development center sponsored by the Office of the Secretary of Defense, the Joint Staff, the Unified Combatant Commands, the Department of the Navy, the Marine Corps, the defense agencies, and the defense Intelligence Community.

Comments are welcome and should be addressed to Christine Eibner at Christine_Eibner@rand.org. For more information on the RAND Health Center for Military Health Policy Research, contact

Susan Hosek at Susan_Hosek@rand.org or Terri Tanielian at Terri_Tanielian@rand.org. A profile of RAND Health, abstracts of its publications, and ordering information can be found at www.rand.org/health. For more information on RAND's Forces and Resources Policy Center, contact the Director, James Hosek. He can be reached by email at James_Hosek@rand.org; by phone at 310-393-0411, extension 7183; or by mail at the RAND Corporation, 1776 Main Street, Santa Monica, California 90407-2138. More information about RAND is available at www.rand.org.

Contents

Figure and Tables

Figure

Tables

Summary

This monograph examines the feasibility of a new model for maintaining the clinical skills of the military medical force. Under the model, active-duty personnel would be assigned to civilian settings during peacetime. The study on which this monograph is based explored the feasibility of this model from a civilian perspective, focusing on civilian receptiveness to the proposed arrangement and identifying potential barriers and concerns. The study found that civilian medical organizations are generally receptive to the idea of such a model and that DoD could consider conducting a pilot study to assess the effectiveness of the model in improving military medical readiness.

Background

Currently, most military medical personnel are stationed at military treatment facilities (MTFs), where they maintain their clinical skills by treating beneficiaries of TRICARE, the military health care program. Yet the medical skills required during deployment are likely to differ significantly from those required in MTFs. Specifically, the most frequent diagnoses during deployment relate to wounds, fractures, and acute conditions such as febrile illness. By contrast, the most frequent diagnoses at MTFs relate to obstetric care and conditions associated with aging.

Consequently, alternative arrangements for maintaining medical skills for deployment may be needed. One alternative would be to station some military medical personnel in nonmilitary settings in which the case mix more closely resembles the expected case mix under deployment, such as emergency rooms or trauma centers. For the model to work, civilian organizations must be willing to accommodate military medical personnel despite the risk of deployment and—preferably—to share the cost with DoD. Would civilian medical organizations accept this partnership?

Study Purpose and Approach

This analysis explored the feasibility of using DoD medical personnel in nonmilitary medical settings, focusing on the civilian perspective. We conducted the analysis in two steps. First, we worked with DoD's Office of Program Analysis and Evaluation to develop a straw-man scenario under which DoD personnel could be stationed at civilian facilities. This model proposed a five-year initial period of service that would follow graduate medical education. Once medical personnel entered military service, they would be stationed at civilian locations on a semipermanent basis. During a typical year, medical personnel would be at their civilian stations for approximately eight months and deployed for day-to-day operations or in military-specific training for four months. Second, we used the straw-man model to gauge civilian reactions to the proposed arrangement. To do this, we interviewed nine civilian health care organizations to determine their willingness to consider the proposed arrangements. The interviews focused on concerns about potential deployment, malpractice liability, cost sharing, compensation, and workforce management issues.

Findings

The analysis found that civilian organizations overall had positive views about accommodating DoD personnel and would be willing to consider sharing the cost of assigning military health care providers to civilian facilities. Civilian organizations felt that the model made the most sense for occupations that perform acute, short-term care. In general, the civilian organizations thought that the model was feasible. However, three reservations about feasibility emerged: (1) if the civilian counterpart job is unionized, the model would be difficult to implement; (2) enlisted DoD medical personnel are occasionally given more responsibility than their civilian counterparts are legally allowed; and (3) the labor market for enlisted medical occupations can be relatively slack. The study results also indicated the following:

- The risk of deployment and liability issues were, somewhat surprisingly, not a major concern.
- Of greater concern were personnel policy issues. In particular, some respondents wondered whether civilian organizations could exert sufficient control over military personnel and accept or reject specific appointments based on organizational needs.
- Three additional questions about personnel policy also arose: Would civilian employers be able to discipline or fire military employees who were not performing adequately? How would legal issues such as sexual harassment and workers' compensation be handled? Would civilian employers have the flexibility to reallocate DoD personnel across geographic locations as needed?
- All the organizations except one (a fire department) expressed a willingness to share the cost of using military medical personnel. There were concerns, however, about the complexity of compensation under dual-payer arrangements. There were also concerns about whether the civilian organizations would need to share the costs of benefits as well as salaries.

Potential Advantages and Disadvantages to DoD

Adding a new category of health care providers—active-duty personnel stationed at civilian facilities—can contribute to readiness. This new category could also increase the military's flexibility by allowing DoD to employ virtually any mix of medical personnel without having to sustain them in MTFs. This flexibility would be useful for employing specialties or maintaining skills that are required for deployment but are seldom used to fulfill DoD's benefits mission. Although reservists can provide this flexibility to a degree, there is no guarantee that the work that reservists are doing in their civilian jobs matches the skills required by DoD during deployment. Further, active-duty personnel stationed in civilian settings could be called up more easily than reservists.

While this increased flexibility represents a benefit, the new category of providers could also increase DoD's costs. If MTF providers currently engaged in beneficiary care are shifted to the civilian sector, DoD would have to expend more resources replacing the care they would otherwise provide to TRICARE beneficiaries, perhaps by turning to the civilian sector. The study found that civilian organizations may be willing to provide permanent-duty stations for military medical personnel and that they may even be willing to cost share for these personnel. If so, this cost sharing would at least partially offset additional costs that DoD might incur under the new model, making it more attractive from DoD's standpoint.

Given the relatively positive reaction of civilian organizations, DoD could consider conducting a pilot study to assess the model's effect on readiness, retention, and morale and to determine whether the benefits of the program appear to outweigh the costs. We anticipate that a meaningful pilot study would involve at least five to seven civilian sites, allowing DoD to have sufficient perspective on the hurdles and contingencies that might arise when negotiating contracts with civilian organizations.

Acknowledgments

We wish to thank several individuals for their guidance and support in carrying out this work. We are especially grateful to John Whitley for his supporting research and for providing valuable information and insights into the study findings. We also thank Steve Carter, COL Tony Carter, Herb Coley, COL Kenneth Cox, CDR David Drozd, Rachel Foster, CDR Dave Tomlinson, CDR Paul Toland, LCDR Christopher Meyer, and participants in the Military Medical Review Working Group for providing comments and suggestions throughout the course of this research. RAND colleagues Andrew Baxter, Donna Farley, Susan Hosek, Jean Jones, Nicole Lurie, Melinda Moore, Janice Nickie-Green, Jeanne Ringel, Terri Tanielian, and Barbara Wynn provided helpful guidance as we initiated this work, developed ideas for the pilot study, and prepared the final draft. Steven Garber and Michael Greenberg, also of RAND, helped us to understand some of the legal issues surrounding military-civilian partnerships. We also thank David Adamson for his editing work on this project and Jennifer Kavanagh for outstanding research assistance.

Finally, we thank numerous anonymous respondents who participated in our interviews with civilian health care organizations. Without their insights, this monograph would not have been possible.

Abbreviations

CBO	U.S. Congressional Budget Office
CRNA	certified registered nurse anesthetist
DNBI	disease and nonbattle injury
DoD	U.S. Department of Defense
FTCA	Federal Tort Claims Act
MTF	military treatment facility
RFA	request for application
VA	U.S. Department of Veterans Affairs
WIA	wounded in action

CHAPTER ONE

Introduction

The U.S. Department of Defense (DoD) is undergoing a significant transformation in response to changes in national security strategy. Among the goals of the transformation are to streamline the armed forces and to improve incentives for individuals to perform their jobs effectively. Possible transformation measures include changing the organization of the forces to allow for greater speed and flexibility in deployment, closing unused or unneeded bases, and ensuring that U.S. troops are stationed in locations that would most effectively maximize their responsiveness and deterrent capacity (National Defense Panel, 1997; Office of Force Transformation, 2004; Feith, 2004).

In light of the ongoing force transformation, DoD asked RAND to assess an alternative paradigm for maintaining the operational readiness of the DoD medical force. Currently, most nondeployed military medical personnel are stationed at military treatment facilities (MTFs), where they maintain their clinical skills by treating TRICARE beneficiaries—active-duty and retired personnel and their dependent family members. Critics of the current system point out that the skills needed to treat this beneficiary population are quite different from the skills required during deployment, necessitating additional training to fully prepare DoD medical personnel for deployment roles (Singer, 1994; CBO, 1995). The primary feature of the new paradigm would be to increase the use of nonmilitary settings for the maintenance of clinical skills.

We note that the idea of relying on the civilian sector to help accomplish DoD's mission is not new, and several previous publications have discussed the possibility of increasing the use of the civilian sector to provide care for TRICARE beneficiaries (Hosek, Bennett, et al., 1995; CBO, 1995). This monograph contributes to the ongoing debate by exploring whether the civilian sector could be used to maintain clinical skills required for readiness.

Specifically, we broadly investigate whether military medical personnel could be stationed at civilian facilities on a day-to-day basis while maintaining their ties to DoD as active-duty service personnel who are therefore accessible for deployment. First, we describe DoD's current approach to medical personnel skill maintenance, followed by a discussion of the changes DoD is considering. We then present results from a targeted set of interviews with civilian health care organizations, assessing the feasibility of the new paradigm from a civilian standpoint. Next, we discuss some of the advantages and disadvantages of this model from DoD's perspective. Finally, we discuss the possibility of implementing a pilot study to more fully evaluate the proposed model.

CHAPTER TWO

Model for Maintaining Military Medical Skills in Civilian Health Care Facilities

Under the current force structure, the military health system serves two primary missions: (1) to provide and maintain readiness for the provision of medical support during military operations (readiness mission) and (2) to provide a comprehensive medical benefit to eligible beneficiaries (benefits mission). TRICARE, the military medical benefit, is a generous insurance policy available to all active-duty personnel, active-duty dependents, retired personnel, and retirees' dependents. TRICARE permits non–active-duty personnel, including retirees, to receive care at MTFs on a space-available basis. The dual nature of the military mission can lead to cost savings because, when not deployed, military medical personnel spend their time treating beneficiaries. However, the dual system also has the potential to create inefficiencies. The mix of medical specialties needed in military operations differs significantly from the skill mix needed for the benefits mission, limiting DoD's ability to seamlessly allocate peacetime medical care providers to wartime roles. Further, deployments can leave MTFs understaffed, at times creating the need to backfill MTFs with reserve and civilian personnel (Cecchine et al., 2001).

Reports by the U.S. Congressional Budget Office (CBO, 1995) and Hosek, Buchanan, and Goldberg (1985) underscore the mismatch between skills needed at the MTFs and skills needed in deployment. The CBO report compares the top 50 principal diagnoses treated at MTFs with the top diagnostic categories among U.S. Marines in Vietnam. Principal diagnostic categories at the MTFs in 1993 were dominated by obstetric care and diseases associated with aging, including coronary atherosclerosis, chest pain, and cataract surgery. In contrast, the most commonly encountered wounded-in-action (WIA) injuries included multiple types of opens wounds and fractures, while the most common diseases and nonbattle injuries (DNBIs) included febrile illness, cellulitis and abscess, and ill-defined conditions. Hosek, Buchanan, and Goldberg (1985) analyzed Air Force inpatient records from 1980 to determine whether the peacetime case mix prepared active-duty physicians to practice for their wartime roles. They found that only 3.6 percent of 273,760 records reported a diagnosis that could be considered war related. Although the numbers in these reports are now outdated, recent conversations with representatives from the Office of the Secretary of Defense and the TRICARE Management Activity confirm that similar disparities exist between the diagnoses currently treated at the MTFs and the conditions treated during deployment in Operation Iraqi Freedom.

The gap between the skills required at the MTFs and the skills required during deployment poses significant challenges to maintaining the operational readiness of the medical force. The literature on health care quality emphasizes the importance of provider case volume as a predictor of patient outcomes (Cowan et al., 2002; Halm, Lee, and Chassin, 2002; Gandjour, Bannenberg, and Lauterbach, 2003; Wang et al., 2004). DoD has attempted to increase medical personnel's exposure to trauma by arranging for forward surgical teams to rotate through civilian trauma centers for training prior to deployment (GAO, 1998). These training programs effectively increase exposure to trauma surgery among the DoD health professionals who have rotated through them (Schreiber et al., 2002). However, the programs are relatively short in duration (two to four weeks), and there are no systems in place to ensure that all trauma personnel rotate through the training centers or that training occurs immediately prior to deployment.[1] Further, the trauma training facilities do not provide experience in the DNBI conditions expected during military operations.

[1] This assessment is based on conversations with supervisory staff at two trauma training centers.

A potential solution to the skill mismatch is to station DoD medical staff at civilian health care organizations in which the case mix is more similar to that expected during deployment. CBO (1995) found that 98 percent of conditions seen at Baltimore's R. Adams Cowley Shock Trauma Unit in 1993 corresponded to WIA conditions expected during wartime, in contrast to less than 10 percent correspondence at MTFs. Part of the University of Maryland Medical Center, R. Adams Cowley is a level I shock trauma center, and correspondence rates are likely to be lower at typical civilian hospitals. However, to the extent that civilian organizations are willing to cooperate, DoD could selectively station medical personnel in facilities (e.g., emergency rooms) where the case mix matches their skill-sustainment needs.

For illustrative purposes, RAND worked with DoD's Office of Program Analysis and Evaluation to develop a straw-man scenario under which DoD personnel could be stationed at civilian facilities. This model proposed a five-year initial period of enlistment or obligation, for which certain types of medical personnel, including physicians and nurses, would be recruited after they complete their medical education. Once medical personnel entered the military, they would be stationed at civilian locations on a semipermanent basis. During a typical year, medical personnel could expect to be at their civilian stations for approximately eight months and deployed for day-to-day operations or in military-specific training for four months. Since training and day-to-day deployments would be predictable, DoD could stagger rotations so that a constant stock of personnel was stationed at the civilian location at all times. For example, there might be 12 personnel assigned to a particular civilian location, with eight available for civilian duty and four engaged in military activities at all times. DoD would have the ability to access all personnel for unforeseen "surge" deployments at any time, perhaps allowing half the staff to stay at the civilian organization for 30 days after receiving orders to ensure a smooth transition.

This straw-man proposal was meant to be illustrative rather than definitive. If actually implemented, the specific details of the model—such as the length of the initial enlistment or obligation period and the amount of time allocated to predictable deployments and training—might vary by service, occupation, and the size of the program. The model is useful primarily as a starting point for thinking about potential partnerships between DoD and civilian organizations.

One of the primary advantages of forging partnerships with the civilian sector is that this model could increase DoD medical personnel's exposure to diagnoses expected during military operations. However, there are other potential advantages as well. Perhaps most important, the use of the civilian sector would give DoD the flexibility to have almost any combination of medical occupations and skill sets without necessarily needing to accommodate these personnel in the MTFs. In the short run, this would provide a systematic way to accommodate personnel from highly specialized fields, such as cardiothoracic surgery, for which demand at MTFs is relatively low. Further, by creating a medical force that is dedicated entirely to the readiness mission, DoD could access these personnel easily and without having to worry about the impact on the benefits mission. In the long run, the model would also allow DoD the flexibility to close or outsource MTFs where beneficiary care is not being provided in a cost-effective manner. To the extent that potential service personnel value the ability to work in the civilian sector, the model might also improve recruitment and retention.

However, assigning military medical personnel to civilian facilities would increase costs unless DoD received some reimbursement or offsetting savings for the care provided by these personnel. There are several ways in which DoD's costs for personnel time spent in civilian facilities might be offset. Civilian organizations may be willing to pay for their share of DoD's personnel costs if the military staff generate patient revenue in the same way that the organization's employees would. The services provided to patients covered by Medicare, Medicaid, and private health plans may be eligible for reimbursement. For the model to be attractive, these reimbursements or other benefits would have to significantly offset the cost savings currently achieved by combining the benefits and readiness missions. Chapter Four presents a more detailed evaluation of the pros and cons from DoD's perspective.

CHAPTER THREE

Feasibility from the Civilian Standpoint

A key feature of the model considered in this monograph is its reliance on the civilian sector to provide permanent duty stations for military medical personnel. DoD would provide medical personnel to civilian organizations at what amounts to a subsidized rate, and, in exchange, DoD would reserve the right to call up these personnel for deployment with no binding restrictions. Because, to implement the model successfully, DoD would need to forge partnerships with the civilian sector, we spoke to representatives from an array of civilian health care organizations to assess the model's feasibility. In total, we spoke to representatives of nine civilian health care organizations, including two civilian trauma centers with existing relationships with DoD through the trauma training program. Other civilian organizations represented in our sample included two regional offices of a group/staff HMO that owns hospitals, a local U.S. Department of Veterans Affairs (VA) medical center, representatives from the VA's Office of the Under Secretary for Health, a professional advocacy association, an academic medical center, and an urban fire department. Although the VA is a government organization that serves veterans, we characterize it as a civilian employer in the context of this monograph because it is staffed by civilian medical personnel and because it is separate from DoD.

We selected these nine sites to get a representative look at the different types of civilian organizations with which DoD might consider partnering. We purposely selected sites that hire a mix of physicians, nurses, and other medical personnel, including technicians and paramedics. In addition, we selected organizations that use salaried physicians (the group/staff HMO and the VA), as well as organizations that grant physicians privileges to work on a fee-for-service basis (the academic medical center and the trauma centers). We interviewed the advocacy group to get a sense of how medical personnel might react to the model. The fire department was selected because it could potentially provide training for enlisted personnel, such as paramedics. Our interviews with the trauma centers were designed to understand the arrangements underpinning existing partnerships between DoD and civilian trauma facilities. The other interviews attempted to assess the overall feasibility of the proposed paradigm from the civilian standpoint and to determine the likely challenges that DoD would have to negotiate to successfully implement partnerships with the civilian sector.

We received approval from RAND's Human Subjects Protection Committee to conduct interviews for this project. As part of this agreement, we keep the identities of interviewees and the names of their employers confidential.

Results from Trauma Training Center Interviews

As part of our assessment, we interviewed civilian and military representatives at two level I trauma centers that have existing relationships with DoD. Specifically, the trauma centers provide training to forward surgical teams that rotate through the centers prior to deployment. DoD training partnerships with civilian trauma centers grew out of research conducted during the early 1990s, suggesting that many military medical personnel received little or no hands-on trauma experience in the context of their peacetime military assignments (GAO, 1992, 1993a, 1993b). To address this issue, the 1996 National Defense Authorization Act (P.L. 104-106) established the Combat Trauma Surgical Committee, which has since facilitated the development of five trauma training programs (one each for the Army and Navy and three for the Air Force) at level I civilian trauma facilities. We spoke with representatives at two of these facilities to understand the negotiation process involved with forming DoD-civilian training partnerships and the facilities' experiences with the partnerships and to assess whether the training programs could be expanded to accommodate a more permanent DoD presence. We conducted interviews using a semistructured protocol, described in Table 3.1.

We learned that the trauma centers typically have 10 to 15 permanent DoD employees whose main function is to facilitate the training program. These staff members are drawn primarily from disciplines related to trauma care, including general surgery, orthopedic surgery,

Table 3.1
Interview Protocol for Trauma Training Centers

Issue	Specific Questions
The role of permanent DoD personnel	We asked respondents to discuss the types of DoD personnel stationed at the trauma centers on a permanent basis and to describe the roles of these personnel. We focused on understanding the occupational composition of permanent DoD personnel, as well as whether permanent personnel treated civilian patients.
The role of trainees	We asked respondents to describe the occupational composition of trainees who rotated through the program, and we asked whether trainees ever treated civilian patients. We also asked about the total number of trainees rotating through the program, as well as the length of time trainees spent on site.
Licensure	We asked what licensure requirements were necessary for permanent personnel and for trainees.
Liability	We asked about how malpractice insurance was handled for permanent staff and for trainees.
Deployments	We asked whether permanent staff were ever deployed, and, if so, we asked about the challenges these deployments posed.
Cultural issues	We asked whether issues related to differences in organizational culture had either hindered or facilitated collaboration with DoD.
Pay and reimbursement	We asked how the salaries of permanent DoD staff and of trainees were paid.
Expanding partnerships	We asked whether DoD personnel would likely benefit from additional training time at the trauma center and whether it would benefit medical readiness to permanently station additional DoD personnel at civilian facilities.

nursing, anesthesiology, and enlisted specialties related to trauma such as operating-room technicians. In addition to facilitating the trauma program, some of these personnel—particularly physicians—treat civilian patients. Trainees typically rotate through the center as part of forward surgical teams, and they are often, but not always, drawn from trauma-oriented backgrounds.

Permanent staff at the trauma facilities must be licensed in the appropriate state. The licensure requirements can be a hurdle because licenses are costly to acquire and it can take time for licenses to be processed. One of our contacts noted that it sometimes takes as long as seven months to get a license instated, which is a sizable wait, considering that the permanent staff at the trauma centers are often reassigned after two years. Physicians practicing in MTFs or in other government facilities, such as VA hospitals, need to be licensed in one state only, regardless of the location of their permanent duty stations.

Although Congress could attempt to extend this licensing arrangement to cover military physicians treating patients in civilian facilities, previous attempts to establish national medical licensing requirements have been called unconstitutional (Herscha, 1996). According to the Federation of State Medical Boards (2006), states are currently working to improve physician mobility and increase license portability. While these changes will not eliminate the need to acquire a new license when moving out of state, the goals are to eliminate paperwork redundancies and reduce the time required to obtain a new license. Ultimately, these changes may make it easier for DoD personnel to obtain state licensure quickly and efficiently. In any case, the model explored here would involve a longer civilian assignment (minimum of five years), and many of the military physicians would be newly trained and could get the state license required by DoD in the state to which they are assigned.

Although medical malpractice is a significant concern at the trauma centers, DoD has been able to address this issue in the context of the trauma training programs. DoD personnel acting under military orders are protected from malpractice lawsuits by the Federal Tort Claims Act (FTCA), which requires the federal government to assume responsibility for any malpractice claims directed at DoD physicians

(P.L. 79-601). One of the two trauma centers in our sample accepts the federal tort protection as sufficient malpractice insurance to cover DoD physicians working at the facility, though they acknowledged that the federal government's "deep pockets" could increase litigants' incentives to sue the hospital. At the second trauma center, federal tort-claim protection was deemed insufficient to fully protect the hospital and its staff from legal accountability. The hospital's primary concern was the "borrowed-servant" doctrine, which allows employers to be held accountable for negligent actions of individuals working under their supervision. Because the hospital believed it was exposed to liability under the borrowed-servant clause, it required DoD to purchase additional malpractice insurance through the affiliated university's malpractice insurance policy. For the permanent trauma center staff, this insurance can cost in the range of $88,000 per physician per year.

Since the trauma centers have a unique arrangement that allows DoD personnel to treat civilian patients, we asked whether DoD deployments had caused significant challenges for them in terms of staffing. Deployment did not pose a significant challenge for the civilian facilities for two reasons. First, deployment of permanent trauma center personnel is relatively rare, in part because the demands of the training programs require all staff to be present. Second, although DoD staff members treat civilian patients, they are stationed at the trauma centers primarily to facilitate the training program, and much of their time is dedicated to DoD-specific work. In some cases, individual DoD personnel have made an effort to become highly involved in civilian care. While the trauma center staff would miss these personnel in the event of deployment, the strain would not pose an overwhelming burden.

Our next set of questions dealt with pay and reimbursement issues for DoD staff members working at the trauma centers. A small number of permanent DoD staff (e.g., on the order of two or three physicians) have privileges to treat civilian patients at the trauma centers. The trainees, who are present for 14 to 34 days, depending on the service, work under civilian supervision for most of their time at the trauma centers. However, trainees are provided several opportunities to take over some or all trauma care responsibilities (Willis, 2003). For example, one of the training programs culminates with a capstone event during which

DoD personnel manage the ER for a period of 36 hours. Because, on average, the trauma center anticipates treating 10 patients a day, this translates into approximately 15 patients treated solely by DoD staff. (Throughout this exercise, however, civilian personnel are available on the sidelines.) Yet while DoD staff treat civilian personnel on some occasions, both permanent DoD staff members and DoD trainees at the trauma centers are fully paid by the military. Typically, the civilian facility can bill for services provided to privately insured patients, but any reimbursements are not passed on to DoD.

Although the trauma centers seemingly gain from being able to bill for some of these patients, the centers view this as a "break-even" arrangement. Section 1814(c) of the Social Security Act (P.L. 74-271) bars government agencies such as DoD from billing for patients insured through Medicare or Medicaid. Each trauma center interpreted the law slightly differently, but both agreed that the legislation restricts the ability to bill for services provided to patients insured through Medicaid or Medicare when the attending physician is a DoD staff member. Because the trauma centers' patient population is disproportionately low-income, the inability to bill Medicaid is a significant drawback. In fact, the memorandum of understanding at one of the centers requires that, because it is impossible to determine the insurance status of trauma patients before administering care, DoD must compensate the center for lost billing that occurs when a DoD physician inadvertently treats a Medicaid or Medicare patient. Another assumption motivating this compensation arrangement is that there is no increase in patient volume due to the DoD presence. Thus, because trauma center physicians are reimbursed on a fee-for-service basis, civilian physicians potentially lose an opportunity to bill for services when DoD staff treat a patient.

Finally, representatives of the trauma centers felt that DoD could benefit from expanding DoD-civilian partnerships either by increasing the length of the training programs or by stationing more military personnel at the centers on a permanent basis. Whether DoD would be able to recoup any of the billing for staff permanently stationed at the trauma centers is unclear. In at least one situation, the university affiliated with the DoD training center nearly terminated the program due to concerns about financial losses that could stem from the DoD presence (in terms of lost billing, overhead, and other expenses), suggesting that participating centers might be resistant to reimbursing DoD. However, as we learned in our other interviews (discussed in the following section), many civilian organizations would consider at least partially reimbursing DoD for personnel costs.

Interviews with Other Civilian Organizations

In addition to our conversations with representatives of the trauma centers, we also conducted interviews with representatives of six civilian organizations that employ medical personnel and one professional advocacy group that sets accreditation standards and lobbies for medical personnel. The intent of these interviews was to determine whether these organizations would consider partnering with DoD to provide permanent duty stations for military medical personnel and, if so, under what conditions. We selected these organizations purposively to get representation from a variety of civilian employers, including a group/staff-model HMO, the VA hospital system, an academic medical center, and an urban fire department. We also spoke to a professional advocacy association to get feedback on how the arrangement would affect employees. Our interviewees came from a variety of disciplinary backgrounds and included physicians, researchers, attorneys, and hospital administrators. Usually, we spoke to several representatives at the same location. For large organizations that had facilities across the country (the VA and the group/staff HMO), we conducted interviews at multiple sites. At the VA, we spoke to both local- and headquarters-level personnel.

We started our interviews by discussing the straw-man model presented in Chapter Two, clarifying that the model was meant to illustrate the type of partnerships DoD is considering but that the details were subject to change. Next, we used a semistructured interview protocol to assess respondents' reactions to the possibility of participating in DoD-civilian health care partnerships. In contrast to our trauma center interviews, which were meant to understand existing partnerships between DoD and the civilian sector, these interviews sought to determine the feasibility of forging new relationships between DoD and the civilian sector. Table 3.2 describes the protocol that we used for the civilian interviews.

With one exception (the fire department), all of the organizations thought that partnerships with DoD would be feasible, conditional on reaching the right agreement. Of the six organizations that hire medical personnel (VA headquarters, the local VA, the two HMO sites, the academic medical center, and the fire department), all but the fire department were willing to consider cost-sharing arrangements with

Table 3.2
Interview Protocol for Other Civilian Organizations

Issue	Specific Questions
Overall feasibility	We asked respondents about their overall reactions to the feasibility of DoD-civilian partnerships.
Major hurdles and benefits	We asked about the major hurdles that might stand in the way of DoD-civilian partnerships. We also asked about any benefits that they foresaw.
Occupations	We asked whether the model were better suited to particular occupations and, if so, which ones.
Licensure	We asked whether respondents had concerns about licensure issues. We asked whether the civilian organizations would be willing to accommodate DoD personnel with out-of-state licenses, conditional on legislation permitting this sort of arrangement.
Liability	We asked whether the civilian organizations had specific liability concerns and whether these concerns differed by occupation.
Deployments	We asked whether the organizations could accommodate personnel who were available for only part of the year. We also asked whether the organizations could cope with unforeseen surge deployments.
Cultural issues	We asked whether there were organizational culture issues that would inhibit collaboration with DoD. We also asked about the main values and goals promoted by the organization and whether the representatives thought that DoD would share these values.
DoD salaries	We asked whether the organization would be willing to pay for all or some of the DoD personnel's salary.

DoD. When asked about hurdles, most of the organizations raised questions about the administrative details that would have to be worked out to make the partnerships successful. Contrary to our expectations, issues related to liability and deployments did not typically emerge as major impediments to DoD-civilian partnerships. Rather, the major questions had to do with personnel policy and compensation issues. Several respondents wondered how much control the civilian organizations would have over who was stationed with them. They asked whether they would be able to request (or reject) specific personnel, whether they could request (or reject) specific occupations, and whether they would be asked to accommodate a "basket" of personnel. Not surprisingly, the model is more attractive to civilian organizations when it allows them more control over who is stationed with them. Although the inability to select specific personnel would not rule out the model, being required to accommodate unneeded personnel would be difficult.

The fire department stood out as the only organization that did not foresee being able to accommodate military personnel on a permanent basis. There were several reasons for this reaction. First, firefighters have a strong union that would undoubtedly protest the influx of discounted labor from DoD. The department was also unconvinced that it could adequately indemnify itself against all potential legal liabilities, such as a DoD employee being hurt on the job. Additionally, the fire department has a substantial surplus of applicants for job openings, and the hiring process is extremely competitive. (For example, our contact department recently had 16 vacancies for which it received 6,000 applications.) Because the labor market for these positions is competitive, partnering with DoD would not benefit the fire department through easier access to hard-to-recruit personnel. Finally, virtually all firefighters are hired at entry level and are given significant training before being fully accepted as firefighters, and it would be difficult to break this model to accommodate DoD personnel. However, despite these barriers to stationing DoD personnel at the fire department on a permanent basis, the fire department currently supports a one-week ride-along program through which DoD personnel accompany firefighters on calls. Our contact, who was an Army combat medic before joining the fire department, felt that the ride-along program was very valuable and could potentially be expanded. However, the ride-along program is considered a training program, and it is paid for by DoD.

Organizations other than the fire department were more optimistic about partnering with DoD. Nevertheless, several issues related to personnel policy arose in our interviews. First, respondents wondered whether they would have the authority to discipline or terminate DoD employees who, in their opinion, were not performing adequately. Although the organizations raised this issue in the context of terminating a single employee with performance problems, a separate concern might relate to the organizations' ability to terminate the entire arrangement with DoD. Civilian organizations also asked about how legal issues such as workers' compensation lawsuits and sexual-harassment claims would be handled and whether DoD-civilian partnerships would expose them to liability in this regard. Finally, the organizations with multiple locations wondered whether they would have the flexibility to reallocate DoD personnel if needed. This could mean occasionally transferring DoD personnel from one geographic area to another to accommodate changes in demand for labor. Alternatively, it could mean using DoD staff in per diem roles, in which personnel stay in the same geographic region but rotate among facilities on a daily or weekly basis, depending on demand.

A second set of questions identified by the civilian organizations had to do with compensation and benefits. One respondent noted that payment systems could be complicated under the proposed model if the salaries of military personnel were paid both by DoD and by the civilian organization. This respondent felt that the model would work best if the paperwork and administrative burden on the civilian organization were minimized; this might mean that the civilian organization would pay DoD a lump sum each year and DoD would distribute paychecks to military personnel. Several organizations asked whether they would be required to provide benefits for DoD personnel in addition to salary. One organization suggested that the model would be very attractive if DoD would pay the full cost of health benefits for military personnel, effectively providing the civilian organization with a substantial discount on labor costs.

The issue of deployment did not pose as much of a hurdle as we expected in terms of civilian organizations' willingness to consider the model. Two civilian organizations noted that they had a number of personnel who were reservists, so they were accustomed to losing personnel temporarily due to military call-ups. Most organizations responded positively to the suggestion that anticipated deployments might be staggered so that a constant stock of personnel would be available to the civilian facility on a regular basis.

However, the possibility of surge deployment would likely affect the way civilian organizations make use of DoD personnel. For example, civilian organizations might rely on DoD surgeons for elective rather than nonelective surgery so that procedures could easily be rescheduled if DoD staff became unavailable. This would reduce the value of the model to DoD if, as seems likely, maintaining military-relevant medical skill involves more nonelective than elective surgery. One organization suggested that it might use DoD staff in per diem—rather than permanent—roles to make the transition easier in the event of a deployment. Another organization suggested staffing DoD personnel in remote locations, where the benefit of additional personnel would outweigh the drawback of deployment. Several organizations said that DoD staff would not be used to fill positions that require long-term provider-patient relationships. Again, at least some of these restrictions might make the partnership less valuable in sustaining the skills critical for military medicine.

Civilian organizations felt that the model made the most sense for occupations that perform acute, short-term care. One respondent mentioned that most surgeons fall into this category. From the civilian standpoint, the model is also well suited to occupations in which staffing is stressed but not in crisis. Occupations that respondents frequently suggested as being compatible with this model included radiologists, urologists, dermatologists, anesthesiologists, cardiologists, orthopedists, nurses, and some technicians. The model is less suited to occupations such as primary care and obstetrics, in which patients expect to see the same provider on a repeated basis. The model would also be challenging to implement for occupations in which the civilian counterpart job is unionized, since unions would likely resist the influx of subsidized competition from DoD. Occupations that fall into this category include enlisted emergency medical technician and paramedic positions, as well as some nurses.

For some occupations, DoD employees have more autonomy than do their civilian counterparts, making it challenging to move these personnel into the civilian sector. There are both legal and cultural aspects to this issue. Some enlisted personnel are expected to perform duties (such as administering sutures) in military settings that may not legally be performed by individuals in comparable occupations in the civilian sector. At the trauma training centers, enlisted occupations such as Army medics (91Ws) and Navy independent-duty corpsmen are permitted to perform these duties only under close supervision. Respondents from the professional advocacy group were also concerned that cultural differences between military and civilian employment would pose barriers. They argued that some DoD personnel, particularly nurses and certified registered nurse anesthetists (CRNAs), are given more deference as officers in the military than is typically afforded in civilian settings. For example, DoD CRNAs can outrank anesthesiologists and are prepared to be the sole anesthesiology provider on a forward surgical team. These cultural issues make military settings more appealing to some military personnel, which could affect their willingness to participate in the proposed model.

In all our interviews, we asked whether differences in organizational culture, values, and mission between the civilian employer and DoD could pose significant barriers to collaboration. One respondent felt that such issues had the potential to be of major importance and felt that partnerships could be successful only if DoD personnel shared the civilian organization's core values—an interest in practicing evidence-based medicine, a willingness to subject oneself to benchmarking and peer review, and a mission-driven approach. But, in most cases, respondents felt that cultural barriers would be minimal and unlikely to pose a major hurdle. Several respondents—notably, representatives from the VA and the group/staff HMO—mentioned that DoD personnel might share some of their key organizational values, such as an orientation toward service. One respondent suggested that organizational collaboration might be easier if DoD personnel blended in with civilians, for example, by wearing civilian clothes and adopting civilian forms of address. One respondent felt that cultural differences could potentially be beneficial and that DoD personnel might enhance their organizations by bringing a fresh perspective.

At the end of our conversations with the six civilian organizations that hired medical personnel, we asked whether they would consider partially or fully reimbursing DoD for personnel costs. In all but one case (the fire department), the organizations said that they would consider paying at least part of the DoD employee's salary. Of course, ultimate willingness to support salaries (or even to participate at all) depends on whether the details discussed in this chapter could be negotiated to the satisfaction of both DoD and the civilian organization. Further, civilian organizations would need a discount or monetary transfer from DoD to compensate them for the risk associated with deployment. Despite these issues, three organizations—the group/staff HMO, the VA, and the academic medical center—suggested that they would be interested in participating in a pilot study.

CHAPTER FOUR

Advantages and Disadvantages from DoD's Perspective

Currently, DoD jointly fulfills its readiness and benefits missions using a variety of personnel—active-duty medical personnel sustained at MTFs, medical reservists, civilian providers employed at MTFs, and private-sector civilian providers who treat military beneficiaries through TRICARE. As illustrated in Figure 4.1, the proposed model adds a new type of personnel to the mix: active-duty personnel sustained in the civilian sector. This new option, shaded in the figure, creates a category of personnel that—like active-duty providers stationed at MTFs and medical reservists—can contribute to the readiness mission. Adding this new option could be advantageous to DoD in at least three ways: (1) improved flexibility, (2) better maintenance of clinical skills, and (3) improved recruitment and retention. We discuss each of these potential benefits in turn.

Advantages

Flexibility

Stationing active-duty personnel at civilian locations could increase the military's flexibility by allowing DoD to employ virtually any mix of medical specialties without having to sustain them in MTFs. This flexibility would be useful for employing specialties or maintaining skills that are required for deployment but are seldom used to fulfill DoD's benefits mission. Although reservists can provide this flexibility to some degree, active-duty personnel stationed in civilian settings could be called up more easily than reservists and would be available

Figure 4.1
The New Paradigm Adds a Provider Type

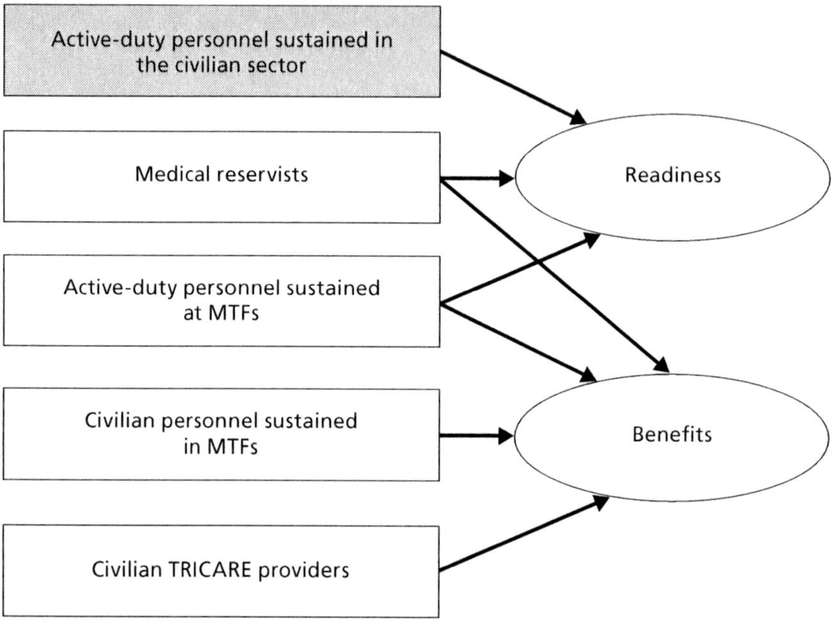

RAND *MG638-4.1*

for ongoing activities during the four months or so that they are on military duty. The model would also add flexibility by allowing DoD to deploy troops without having to make contingency arrangements to sustain the benefits mission. In essence, civilian organizations, rather than MTFs, would be accepting the risk of losing personnel due to deployment. However, since military medical personnel would be spread across many civilian organizations, the risk shouldered by each organization would be low.

Clinical Skills

Civilian organizations could potentially provide DoD personnel with opportunities to maintain clinical skills in fields in which demand at the MTFs is low. We anticipate that civilian experience would be particularly valuable for treating conditions that are common in deployment but unusual at the MTFs, such as traumatic injuries and burns. Previous research (Hosek, Buchanan, and Goldberg, 1985; CBO, 1995) finds that the case mix at MTFs does not closely match the case mix expected during deployment and that civilian hospitals can potentially provide a better fit. However, this evidence is more than 10 years old. To fully understand the current case-mix mismatch problem, it would be necessary to revisit this issue using new data.

Recruitment and Retention

Finally, some of these active-duty personnel would be retained and move up in rank. To the extent that DoD personnel value the opportunity to work in civilian settings, the proposed model could improve recruitment and retention. Military personnel might particularly value the opportunity to work in a civilian setting if their spouses have better career options near the civilian worksite or if experience in the civilian sector improves their ability to transition into civilian employment after leaving DoD. Other military personnel might prefer to work in the MTFs. To a degree, DoD might be able to assign military personnel to the civilian sector based on preference. However, if staff were assigned to the civilian sector against their wishes, retention and recruitment might not improve.

While some of these benefits would be valuable to DoD, adding the new option could increase DoD's costs. If MTF providers currently engaged in beneficiary care are moved to the civilian sector, DoD will have to expend additional resources to purchase beneficiary care, perhaps also by utilizing the civilian sector. In essence, the model eliminates the savings that DoD currently realizes by using the same personnel to fill both benefits and readiness roles. Our interviews found that civilian organizations are potentially willing to provide permanent duty stations for military medical personnel and that they may even be willing to subsidize the cost of these personnel. However, it is unclear whether the savings achieved through civilian cost sharing would outweigh the new costs incurred by splitting the benefits and the readiness missions.

Disadvantages

In addition to the issue of monetary costs, our interviews also revealed some potential drawbacks associated with DoD-civilian partnerships. Our interviews were designed to understand civilian organizations' willingness to collaborate with DoD and get feedback on what would make the model most attractive from their perspective. As a result, our interviews provide insight into the types of arrangements that would be most desirable from the civilian standpoint. In general, the organizations were willing to negotiate, and—although they discussed their most preferred arrangements—they seemed prepared to accommodate other scenarios as well. Nevertheless, in several cases, the civilian organizations expressed a desire to utilize DoD personnel in ways that may not completely align with DoD's needs. Next, we discuss several of these issues

Staffing in Remote Locations

From the civilian sector's perspective, one of the main benefits of this model is that DoD personnel could help fill staffing shortages. Representatives of at least one of the organizations in our study emphasized that they have particular difficultly staffing their rural facilities and that DoD personnel could be especially useful in these settings. Yet, it is not clear that rural facilities would provide experience or case mix that would be relevant for fulfilling the DoD's readiness mission. As an alternative, DoD might attempt to forge partnerships in underserved areas, such as inner cities, where case mix and experience requirements might be more aligned with military needs.

Elective Rather Than Nonelective Surgeries

Representatives of at least two organizations suggested that they would use DoD personnel for elective procedures rather than nonelective surgeries as a way to insulate themselves from staffing crises in the event of a deployment. Yet nonelective surgeries, particularly trauma and emergency surgeries, are likelier to approximate requirements on the battlefield than are other types of surgery. Further, some elective surgeries, such as knee replacements, are frequently performed at MTFs.

In negotiation processes with civilian facilities, DoD would want to explicitly specify the type of clinical experience it expects personnel to receive at the civilian location. If the civilian location is unable or unwilling to provide adequate exposure to the required case mix, DoD might not want to pursue a partnership with the organization.

Per Diem Work

One of the organizations discussed the possibility of using DoD personnel to fill per diem roles, which would require DoD staff to move from location to location within a geographic region on a daily or weekly basis. It might be very difficult to anticipate what the case mix would be under a per diem arrangement, potentially raising the same concerns that rural or elective assignments would raise. Further, this arrangement could be frustrating for DoD personnel due to longer commutes, lack of a permanent workplace, reduced opportunity to form relationships with colleagues, and—possibly—lower status at the civilian organization. In some cases, the clinical experience afforded by the civilian location may be valuable enough to justify the drawbacks of per diem work. Further, some DoD personnel may prefer to work in civilian settings, making them likelier to agree to per diem employment. However, this option likely would be unattractive to many personnel. DoD might consider allowing personnel to refuse a civilian station that required per diem work, unless the clinical experience were unavailable in other settings.

Occupation Mix

Finally, some of the occupations in which the civilian organizations are most interested, such as radiology and cardiology, may be poor matches for this model from DoD's perspective. From DoD's standpoint, this model is best suited for occupations that are required for readiness but cannot be appropriately staffed at MTFs. DoD might find it difficult to appropriately staff certain occupations at MTFs because the occupations are not needed to support the benefits mission or because the case mix seen by these providers at MTFs is not reflective of the case mix expected in deployment.

Service sizing models can give partial insight into which occupations would be suitable for this model. According to a recent CNA Corporation report (Levy, Christensen, and Asamoah, 2006), readiness requirements in fiscal year 2004 exceeded end strength in three physician specialties: anesthesiology, general surgery, and orthopedic surgery. To the extent that the gap between end strength and requirements stems from a lack of work at the MTFs for these specialties during peacetime, DoD could use civilian settings to maintain the required numbers (though, shortfalls in end strength may reflect supply constraints, as well as low demand at the MTFs). For other specialties, the case mix seen at MTFs may not reflect case mix expected during wartime. For example, plastic surgeons stationed at MTFs may work primarily on cosmetic procedures, while the skills they need in deployment primarily involve reconstructive surgery. These case-mix mismatches cannot be identified with the sizing model and require a more careful evaluation of the similarities and differences between procedures performed at the MTFs and in theater, as well as consideration of whether the case mix seen in civilian settings represents an improvement over the case mix seen at MTFs.

CHAPTER FIVE

Implementing a Pilot Study

The proposed model could contribute to DoD's mission by improving its ability to accommodate specialties used infrequently in beneficiary care by providing exposure to a specialized case mix and by reducing the demands placed on MTFs during deployment. Since our interviews determined that civilian organizations are willing to consider accommodating military personnel despite the risk of deployment, DoD may want to test this model through a pilot study. Although we have enumerated several potential benefits of the new paradigm, we do not yet have enough information to determine whether the advantages would outweigh the cost of separating the benefits and readiness missions for some occupations. A small-scale pilot study (e.g., five to seven sites) would shed light on some of these issues. In particular, a pilot study would determine whether DoD could negotiate mutually agreeable contracts with civilian organizations. It would also provide a realistic assessment of the degree of cost sharing to which civilian organizations would be willing to contribute. A carefully designed pilot that allowed DoD to follow medical personnel over time would provide insight into the model's impact on retention and morale—both important considerations, given that, for some personnel, the ability to work in a civilian setting could be attractive. Testing this model through a pilot program would also give DoD a realistic sense of the potential hurdles imposed by state licensing requirements, public and private reimbursement limitations, and other legal barriers.

As part of RAND's evaluation, we organized a roundtable discussion with in-house experts to get insights into how DoD might conduct a pilot for this model. Roundtable participants included two senior health-policy analysts and one senior physician, all with significant experience working with DoD and conducting pilot studies. The consensus among discussion participants was that DoD should consider conducting a pilot study in two phases: a planning phase and a roll-out phase. One of the chief components of the planning phase (phase I) would be a competency assessment to determine which occupations are most suitable for the model. Through the competency assessment, DoD would need to articulate the profile of clinical cases to which it wants its medical staff exposed and the skills and competencies needed to treat these conditions. DoD would then need to map these skills and competencies into occupations. For example, a competency requirement in repairing a fractured leg would be mapped to the occupation of orthopedic surgeon. For each occupation, DoD would then develop a list of formal criteria for assessing competencies in specific areas, based both on clinical standards developed in the civilian arena and on expert judgment of DoD medical staff. These criteria should reflect both the desired case mix and the frequency of exposure needed to keep skills sharp. Once the criteria were established, DoD would need to determine which occupations were getting adequate exposure at the MTFs. Those without adequate exposure at the MTFs would be good candidates for the pilot study. In addition to developing criteria for assessing particular competencies, DoD might want to develop a prioritized list of broader goals for evaluating the success of the pilot. In addition to improvements in the identified competencies, DoD might wish to consider the pilot's effect on morale, retention, and deployments. To limit the scope of the pilot, DoD might consider focusing on medical occupations in the Army only or in the Army and the portion of the Navy that supports the Marines.

The second step in the planning phase would be to determine how much time DoD staff could reasonably expect to spend in civilian organizations, given military-specific time requirements such as deployment and training. In our interviews with civilian organizations, we used an illustrative example in which we stated that military personnel would be stationed at civilian locations for an average of eight months out of every year, with an initial obligation or enlistment period of five years. For each occupation targeted in the model, DoD would need to get a more accurate assessment of the likely schedule. Such information could be determined by speaking to human-resource planners in each represented service.

During the planning phase, DoD would need to determine which civilian organizations are a good match for the model in terms of case mix, interest in the model, and compatibility with DoD's scheduling constraints. Based on our interviews with civilian organizations, discussed earlier in this monograph, we found that one of the main benefits to the civilian participants would be that the model would allow them to access qualified personnel at a lower cost. Our pilot-study roundtable participants uniformly thought this feature made the model particularly attractive to "safety-net" hospitals. Although there are no universally accepted criteria for identifying safety-net hospitals, two key characteristics include a legal or mandated mission to provide care to uninsured populations and a high proportion of care provided to vulnerable patients, such as low-income or uninsured individuals (Gaskin and Hadley, 1999).

Another key feature of safety-net hospitals is that they often provide a substantial amount of care through the emergency department. For example, Baxter and Mechanic (1997) use the percentage of emergency-department care provided as an indicator of a hospital's safety-net status. To the extent that safety-net hospitals provide a disproportionate share of services to trauma and ER patients, targeting these organizations could be mutually beneficial from the perspective of both DoD and the hospital. DoD could benefit because the safety net provides exposure to a specialized type of case mix, and the hospital could benefit from having subsidized personnel. Focusing on safety-net hospitals would make the proposed model quite similar to the National Health Service Corps, through which clinicians are offered loan repayment in exchange for working in underserved communities. A drawback, however, of targeting safety-net hospitals is that these hospitals may have a limited ability to cost share with DoD, particularly if a large portion of their patients are insured through Medicaid.

More generally, civilian sites should be able to both provide exposure to the required case mix and accommodate DoD's scheduling needs. As an additional requirement, DoD might consider the civilian site's compatibility with DoD's organizational culture. For example, DoD might prefer that staff work in environments in which teamwork is valued. Our interviews with civilian sites suggested that a wide range of civilian organizations might consider participating in a pilot study. DoD could formally solicit interest from civilian organizations by sending out a request for application (RFA) summarizing the proposed pilot study and the parameters of the model in terms of targeted occupations, case-mix needs, scheduling, and cost sharing. Interested organizations could respond to the RFA, and DoD could then determine whether an organization is a good match. If no suitable matches were identified, DoD could modify the RFA in an attempt to attract a more suitable array of civilian prospects.

Issues related to licensure and credentialing might pose a challenge for the pilot because participants would need to be licensed in the state in which they practice. DoD could attempt to minimize the burden of licensure requirements by targeting civilian sites in states in which DoD medical staff are likeliest to be licensed. DoD could also conduct an assessment to determine which states have the fewest constraints in terms of licensure and credentialing requirements and then prioritize these locations when recruiting civilian organizations for the pilot. VA sites could also be prioritized for specialties in which the VA case mix is appropriate, since physicians in VA hospitals need only be licensed in a single state. If the model were to be implemented on a larger scale, DoD could consider asking for congressional intervention to allow DoD medical staff to practice in any state so long as they have a license in at least one state.[1]

In the roll-out phase of the pilot (phase II), DoD would enter negotiations with interested organizations to develop contracts. These contracts would need to address the issues identified in our civilian interviews, including expectations regarding the number and type of DoD staff assigned to each location, responsibilities regarding who would have authority to discipline or terminate DoD staff members,

---

[1] Because such an intervention could be seen as infringing on states' rights, it might be challenged on constitutional grounds.

expectations about the case mix and per diem work, and malpractice concerns. The contracts would also need to specify the conditions under which either DoD or the civilian organization could terminate the entire arrangement. Based on the judgment of participants in our roundtable discussion, DoD would likely need to establish contracts with at least five to seven civilian locations for the pilot to yield valuable results. DoD would also have to determine the length of the pilot study. This could be informed by the scheduling requirements identified in phase I of the pilot. For example, the duration of the pilot study might coincide with the duration of the initial obligation period for pilot participants.

Several of our roundtable participants felt that DoD should consider fielding two simultaneous pilots: one for physicians and one for nurses. In theory, the model could apply to enlisted personnel as well, but our interviews with civilian organizations identified several stumbling blocks for nonofficers, suggesting that it makes sense to focus on doctors and nurses first. Looking at doctors and nurses separately might be informative because some of the issues that are likely to affect the success of the program are very different for each of the two groups. For example, there are powerful nursing unions in some parts of the country that may inhibit DoD's ability to negotiate with civilian organizations. Further, hospital-based nurses are almost always directly employed by the facility, while, in many cases, doctors work independently and have privileges to practice outside of certain hospitals. Should DoD decide to field two simultaneous pilots, the minimum number of sites would have to be increased accordingly.

The model that DoD is considering likely requires that staff be salaried, rather than working independently and billing on a fee-for-service basis. The VA holds considerable potential for DoD partnerships because physicians are salaried, licensing is not an issue, and funds could be transferred via interagency agreement without the need to bill third-party payers. Further, there is precedent for VA and DoD sharing arrangements in which physicians from one organization see patients in the other organization's facilities.[2] However, it is not clear whether the case mix at VA hospitals is a good fit for DoD. Large group/staff-model HMOs also hold potential because they employ salaried doctors, but there are few such organizations currently operating in the U.S. health care system. And, like the VA, it is unclear whether the case mix they accommodate would meet DoD's needs. For example, although the group/staff HMO employs surgeons, it does not provide level I trauma care. While physicians often work in traditional civilian hospitals on a fee-for-service basis, there is a growing trend toward using "hospitalists"—doctors employed by hospitals who treat and admit patients (Pham et al., 2005). In addition, academic medical centers may directly employ physicians, and hospitals sometimes employ physicians in ERs (particularly in safety-net hospitals). So, while there are challenges to implementing this model in a traditional civilian hospital setting, there are also opportunities.

For the pilot to be successful, DoD would have to recruit medical personnel to participate in the study. It would probably want to offer medical personnel the choice to participate (rather than randomizing personnel into the trial), which could complicate the subsequent evaluation if particular types of employees self-select into the treatment group. Prior work (see, e.g., Rostker, 2006, Chapter Seven) has addressed this type of selection by making programs available only in certain geographic locations, thus enabling researchers to compare the outcomes of those with access to the pilot to the outcomes of those outside the scope of the pilot. For some specialties in which DoD is not currently meeting recruiting goals, it may wish to include recruitment bonuses, as proposed by Levy, Christensen, and Asamoah (2006), as part of the pilot.

The success of the pilot could be evaluated using some of the occupation-specific criteria developed for the competency assessment conducted in phase I. The criteria would articulate both the types of clinical cases to which DoD medical staff should be exposed and the frequency of exposure required, allowing DoD to compare these goals

---

[2] See President's Task Force to Improve Health Care Delivery for Our Nation's Veterans (2003) for more information on these arrangements.

with the actual experience attained through the pilot. In addition to evaluating the pilot in terms of meeting these specific criteria, DoD could also consider the more general list of goals articulated during the planning phase. Information on issues such as morale could be collected by administering a survey to pilot participants (and to a set of nonparticipating control subjects) during and after their obligation periods. The survey could ask questions about job satisfaction, willingness to recommend military service to others, perceived marketability of skills outside of DoD, intention to continue DoD service, and whether the availability of the civilian program affects their intention to continue service. After the initial obligation ends, DoD could compare the retention rates of pilot participants and those of nonparticipants. Finally, DoD could evaluate whether the availability of the pilot program affected recruitment.

CHAPTER SIX

Summary and Conclusion

This monograph investigated a new model for maintaining the clinical skills of the military medical force, in which active-duty medical personnel would be sustained in civilian settings during peacetime. The model would give DoD medical staff the opportunity to treat a very targeted mix of clinical cases, a feature that would be particularly valuable for specialties that do not get adequate case-mix exposure at the MTFs. In addition, the model would give DoD substantial flexibility to tailor the size and occupational composition of its medical force. Yet, for the model to work, civilian organizations must be willing to accommodate military medical personnel despite the risk of deployment and—preferably—to share the cost with DoD. As part of our research, we conducted interviews with nine purposively selected civilian health care organizations to determine their willingness to consider the proposed model. In general, the civilian organizations thought that the model was feasible, though they expressed hesitancy when considering the model for enlisted specialties. This reluctance stemmed from three concerns: (1) if the civilian counterpart job is unionized, the model would be difficult to implement; (2) enlisted DoD medical personnel are occasionally given more responsibility than their civilian counterparts are legally allowed; and (3) the civilian labor market for some enlisted jobs is relatively competitive, lessening the appeal of DoD partnerships from civilian employers' perspective.

Despite several concerns directed at enlisted personnel, the overall responses of our civilian respondents tended to be optimistic. Most organizations thought that DoD could reasonably expect civilian organizations to share costs, and several organizations expressed interest in participating in a pilot study. Yet our interviews raised questions related to the specific details of the contractual arrangements between DoD and participating civilian organizations. Many of these questions related to the degree of control that the civilian organization would have over choosing, disciplining, terminating, and reallocating DoD staff. Additionally, the civilian organizations raised questions related to legal issues, such as whether DoD staff would be able to sue for workplace-related claims (e.g., workers' compensation) and whether DoD would be willing to purchase malpractice insurance despite protection offered under FTCA (P.L. 79-601). These questions suggest that effective civilian-DoD partnerships would require very thoughtful contractual agreements that anticipate a variety of possible contingencies.

Our interviews also found that, in some cases, civilian organizations may hope to use military medical personnel in ways that do not align with DoD's needs. In particular, several organizations suggested that they might assign DoD staff to remote locations and use them for elective surgeries or per diem work. Nevertheless, the majority of civilian respondents indicated that they were flexible and could negotiate with DoD about specifics such as assignment location and job requirements. To fully assess whether mutually beneficial and sustainable contracts could be achieved, DoD would likely need to develop a pilot study of the proposed model. In addition to providing a testing ground for developing contracts, the pilot study would allow DoD to determine whether employment in the civilian sector improves the clinical skills of selected occupations. Further, a pilot study could follow DoD participants, using both survey data and administrative records to assess the program's effect on recruiting, morale, retention, and willingness to recommend DoD service to others.

References

Baxter, R. J., and R. E. Mechanic, "The Status of Local Health Care Safety Nets," *Health Affairs*, Vol. 16, No. 4, July–August 1997, pp. 7–23.

CBO—*see* U.S. Congressional Budget Office.

Cecchine, Gary, David E. Johnson, John R. Bondanella, J. Michael Polich, and Jerry M. Sollinger, *Army Medical Strategy: Issues for the Future*, Santa Monica, Calif.: RAND Corporation, IP-208-A, 2001. As of September 26, 2007: http://www.rand.org/pubs/issue_papers/IP208/

Cowan, John A., Justin B. Dimick, B. Gregory Thompson, James C. Stanley, and Gilbert R. Upchurch, "Surgeon Volume as an Indicator of Outcomes After Carotid Endarterectomy: An Effect Independent of Specialty Practice or Hospital Volume," *Journal of the American College of Surgeons*, Vol. 195, No. 6, December 2002, pp. 814–821.

Federation of State Medical Boards, "Trends in Physician Regulation," Dallas, Tex., April 2006. As of May 8, 2006: http://www.fsmb.org/pdf/PUB_FSMB_Trends_in_Physician_Regulation_2006.pdf

Feith, Douglas J., Under Secretary of Defense for Policy, "Transformation and Security Cooperation," remarks delivered at the National Press Club Conference, Washington, D.C., September 8, 2004. As of September 26, 2007: http://www.defenselink.mil/policy/sections/public_statements/speeches/archive/former_usdp/feith/2004/september_8_04.html

Gandjour, Afschin, Angelika Bannenberg, and Karl W. Lauterbach, "Threshold Volumes Associated with Higher Survival in Health Care: A Systematic Review," *Medical Care*, Vol. 41, No. 10, October 2003, pp. 1,129–1,141.

GAO—*see* U.S. General Accounting Office.

Gaskin, Darrell J., and Jack Hadley, "Population Characteristics of Markets of Safety-Net and Non–Safety-Net Hospitals," *Journal of Urban Health*, Vol. 76, No. 3, September 1999, pp. 351–370.

Halm, Ethan A., Clara Lee, and Mark R. Chassin, "Is Volume Related to Outcome in Health Care? A Systematic Review and Methodologic Critique of the Literature," *Annals of Internal Medicine*, Vol. 137, No. 6, September 17, 2002, pp. 511–520.

Herscha, Lynnette A., "Is There a Doctor in the House? Licensing and Malpractice Issues Involved in Telemedicine," *Boston University Journal of Science and Technology Law*, Vol. 2, No. 8, April 1996.

Hosek, Susan D., Bruce Bennett, Joan L. Buchanan, M. Susan Marquis, Kimberly A. McGuigan, Janet M. Hanley, Rodger Madison, Afshin Rastegar, and Jennifer Hawes-Dawson, *The Demand for Military Health Care: Supporting Research for a Comprehensive Study of the Military Health Care System*, Santa Monica, Calif.: RAND Corporation, MR-407-1-OSD, 1995. As of September 26, 2007:
http://www.rand.org/pubs/monograph_reports/MR407-1/

Hosek, Susan D., Joan L. Buchanan, and George A. Goldberg, *Reconciling Air Force Physicians' Peacetime and Wartime Capabilities: Demonstration of a Workforce Design Methodology*, Santa Monica, Calif.: RAND Corporation, R-3202-AF, 1985. As of September 26, 2007:
http://www.rand.org/pubs/reports/R3202/

Levy, R. A., E. W. Christensen, and S. Asamoah, *Raising the Bonus and the Prospects for DoD's Attracting Fully Trained Medical Personnel*, Alexandria, Va.: CNA Corporation, CRM D0013237.A2/Final, 2006.

National Defense Panel, *Transforming Defense: National Security in the 21st Century*, report to Secretary of Defense William Cohen, December 1997. As of September 26, 2007:
http://www.dtic.mil/ndp/FullDoc2.pdf

Office of Force Transformation, *Elements of Defense Transformation*, Washington, D.C., October 2004. As of May 30, 2006:
http://www.oft.osd.mil/library/library_files/document_383_ElementsOfTransformation_LR.pdf

Pham, Hoangmai H., Kelly J. Devers, Sylvia Kuo, and Robert Berenson, "Health Care Market Trends and the Evolution of Hospitalist Use and Roles," *Journal of General Internal Medicine*, Vol. 20, No. 2, February 2005, pp. 101–107.

P.L.—*see* Public Law.

President's Task Force to Improve Health Care Delivery for Our Nation's Veterans, *Final Report*, Washington, D.C., May 2003. As of September 26, 2007:
http://veterans.senate.gov/documents/ihcdfinal.pdf

Public Law 74-271, Social Security Act, August 14, 1935.

Public Law 79-601, Federal Tort Claims Act, August 2, 1946.

Public Law 104-106, National Defense Authorization Act, February 10, 1996.

Rostker, Bernard D., *I Want You! The Evolution of the All-Volunteer Force*, Santa Monica, Calif.: RAND Corporation, MG-265-RC, 2006. As of September 26, 2007:
http://www.rand.org/pubs/monographs/MG265/

Schreiber, Martin A., John B. Holcomb, Cass W. Conaway, Kyle D. Campbell, Matthew Wall, and Kenneth L. Mattox, "Military Trauma Training Performed in a Civilian Trauma Center," *Journal of Surgical Research*, Vol. 104, No. 1, May 2002, pp. 8–14.

Singer, Neil M., Acting Assistant Director, National Security Division, Congressional Budget Office, statement on reforming the military health care system before the Subcommittee on Military Forces and Personnel, Committee on Armed Services, U.S. House of Representatives, April 19, 1994. As of September 26, 2007:
http://www.cbo.gov/ftpdoc.cfm?index=4844&type=0

U.S. Congressional Budget Office, *Restructuring Military Medical Care*, Washington, D.C., July 1995. As of July 17, 2007:
http://www.cbo.gov/ftpdoc.cfm?index=5309&type=0

U.S. General Accounting Office, *Operation Desert Storm: Full Army Medical Capability Not Achieved*, Washington, D.C., GAO-NSIAD-92-175, August 1992.

———, *Operation Desert Storm: Improvements Required in the Navy's Wartime Medical Program*, Washington, D.C., GAO-NSIAD-93-189, July 1993a. As of September 26, 2007:
http://archive.gao.gov/t2pbat5/149621.pdf

———, *Operation Desert Storm: Problems with Air Force Medical Readiness*, Washington, D.C., GAO-NSIAD-94-58, December 1993b. As of September 26, 2007:
http://archive.gao.gov/t2pbat4/150526.pdf

———, *Medical Readiness: Efforts Are Underway for DoD Training in Civilian Trauma Centers*, Washington, D.C., GAO/NSIAD-98-75, April 1998. As of September 26, 2007:
http://www.gao.gov/archive/1998/ns98075.pdf

Wang, Henry E., Samuel R. Seitz, David Hostler, and Donald M. Yealy, "Defining the Learning Curve for Paramedic Student Endotracheal Intubation," *Prehospital Emergency Care*, Vol. 9, No. 2, April 2004, pp. 156–162.

Willis, Sharon, "Trauma Training: From the City to the Battlefield," *USU Medicine*, Spring 2003, pp. 6–13.